The Game Through Glass

Playing your youth sports tournament on social media

by Gerard McLean

SharktoothPress

Sharktooth Press

Published in the United States by Sharktooth Press, an imprint of Rivershark Inc. TourneyCentral is a brand of Rivershark Inc. Logomarks registered trademarks of Rivershark Inc.

ISBN: 978-0-9909507-5-2

www.sharktoothpress.com

Cover design: Gerard McLean

Updated for 2017

v1.0_r4

Contents

Things change rapidly in the social media world. Be sure to check out thegamethroughglass.com for errata, marginalia and revisions.

Open the door

The front door

One of my personal frustrations is being given or sold a guide on how to do something, but the writer forgot to mention where the front door is or they left out a step, leaving the reader hanging, wondering where it all fell apart.

I won't do that to you. At least, I don't intend to. This page is your front door and your floor plan. You may be excited to start, but please don't skip around.

This book will get you up and running on the *very basic use of social media tools* for your sports tournament. Once you grasp the basics, how you extend this knowledge and skill set will be your durable competitive advantage for your youth sports tournament.

The first part will explain what social media is and why you need to be integrating it into your sports event.

The second part will guide you through setting up the basic tools you will need to make social media a part of your event, i.e., how to set up a Facebook page, a Twitter account, etc.

The third part will explain how to use the tools together as a cohesive event.

The fourth part will offer some planning and tracking tools so your team and accounts can work together as a unit.

The fifth part is our ongoing social media series on our web site as well as regular webinars and training sessions. Details for this will be managed on www.thegamethroughglass.com

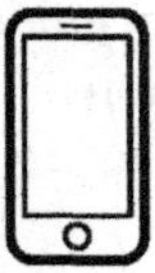

Along the way, you will see an icon of a mobile phone. This means that everything in the section afterwards is intended to be used on a mobile device. Eventually, everything will be mobile so we recommend pushing to do everything on mobile when possible.

We only present tools that are on iOS AND Android, so you never have to choose based on platform. Use what you know.

We write extensively about tournaments. We write even more about using the latest social media tools to host a tournament. The library of documents can be found at our blog called *Touchline*

tourneycentral.com/blog

We also have a regular newsletter sent out at irregular intervals. You can subscribe here:

tourneycentral.com/newsletter-subscription

And of course, the website for this book which has all of that in one place:

thegamethroughglass.com

Let's get started.

Part One

Getting started

What is social media?

Social media is a huge cloud of things like blogging, tweets, likes and status updates so it is easy to become overwhelmed and do nothing.

But you need to roll up your sleeves get in there. We'll help you along the way with this book and our online companion series "social media for sports tournaments."

thegamethroughglass.com

Social media is becoming integrated into the online experience for most people, so if you have not learned to navigate these waters, you may be risking the reach of your tournament. Social media channels can help you reach coaches and teams where they live and help you keep them updated and interested in your event.

Not a game for kids

Do not assume social media channels are for kids. The largest and fastest growing demographic for Facebook and Twitter is age 35-55. That means a player's mom, dad or coach is more likely to "like" you on Facebook or "follow" you on Twitter. While most kids over 13 yrs old will have social media accounts, they connect with their friends, not with brands. If they happen to take a cool photo at your tournament, they may or may not post it. But if they do post it, they will post on their profile, not your tournament page. It's nothing personal; it's just how kids use social media. Their parents on the other hand, may be more likely to connect to your social media channels to remain in the know for news, schedule changes and scores.

Online location is everything

Most websites in general have *wayyy* too many pages! A website today is not the sum total of everything your tournament is online — a very different way of looking at websites from even a few years ago.

The website needs to give the **who what where when why** and be the **authoritative** voice for the your tournament; also, for operationally critical data (sponsors, applications, schedules. The other media stuff needs to live on the outposts like Facebook, Twitter, Instagram, Periscope, Snapchat etc with the website simply validating the authenticity of the content. The outposts are where people live now, wanting your tournament to meld into the rest of their lives, not the other way around.

Your online schedule is a hook for media where you can hang photos, videos, profiles, blog posts, etc. A schedule immediately gives your media context, i.e., who, where, when, why. Use that organic context to your advantage.

The absolute hardest part of shifting to social media channels is convincing folks to let go of about 90% of the website pages. You just don't need them!

Social = Mobile

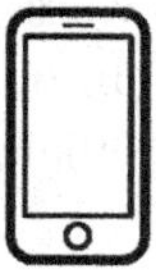

Your tournament is now mobile and real time. Your participants, their fans and supporting community is not waiting until they get home to look up scores, searching for photos, etc. They are doing that in real time as they experience your event.

Because social = mobile, we will be staying on the mobile apps as much as possible, even signing up for accounts using them. For some — like Instagram and Snapchat — there really is no web site; everything is done through the app.

Social media channels priority

Running a tournament is all about making the best choice for what works for you. While it would be great to be everywhere on every social media channel, it's probably not possible.

Here is the list of social media channels in the order of priority that I think you should be working. In a year, this list may change but going into the 2017 tournament season, these are it.

1. Facebook
2. Twitter
3. Instagram
4. Snapchat
5. Periscope
6. YouTube
7. GooglePlus

At minimum, the modern tournament should be fluent in the top four channels to be visible. How much you want to work each channel depends on the manpower you can direct into each. Some of the content to these channels can be automated, like scores and news updates.

Important: Implement the social media tools in order and completely. For example, don't try using Periscope if you haven't yet set up a Twitter account. It will frustrate and scatter you.

Don't get so hung up on the volume of people using specific social media channels. It doesn't matter if one billion people are using Facebook if only your guest teams and local community are willing to connect to your tournament there.

Focus on building your networks where it makes most sense to you. It is better to do fewer channels more deeply than to be everywhere but anemic and unfocused.

Broadcast and interact like everyone in the world is watching you. If you are doing it well, local print and television will ask if they can use your content. If you are doing it exceptionally well, national media will.

Extending reach

There are also other channels that you can explore like Tumblr, Reddit, YouNow, YikYak, FourSquare, Digg, WeChat, etc. But these eight provide a strong core of social media for your soccer tournament. Keep in mind that each channel requires more time and effort to maintain and you should only start with more if you are committed to keeping up these accounts.

The 90 Minute Attention Span

At TourneyCentral, we like to refer to the time your teams are focused on your tournament as the "90 Minute Attention Span." This isn't a good or bad thing, it is just a way of saying that players, coaches and their families care about your event while they are there. Before that, they are busy with their lives. When they leave, they will again be busy with their lives. Your tournament may not ever enter their field of vision until next year.

The goal is to expand their 90 Minute Attention Span by a few seconds or even a full minute. Extend your reach into the social media space by giving players and their families a place to "hook" into you as they stream their lives. If they are at your tournament, they should be tweeting "Having a great time at @yourevent" instead of saying "at a soccer tournament this weekend." If you have a Twitter account, you get named. If not, you don't.

Be on social media but make sure your website remains authoritative. It can get easy to get caught up in the hype about "500 million users are on Facebook" and forget why YOU are there.

Make the social media channels a serious part of your marketing and engagement strategy. Do not delegate it off to a "kid in the club." It is far too valuable. Learn social media as seriously as you would learn how to make a schedule.

Prep work
As with most things, good solid prep work makes the job easier. Do these seven things before attempting to set up accounts:

Do each in order; Don't skip any step.

1. Make a SQUARE 500px x 500px logo (avatar) for your tournament. Make sure it is consistent with your tournament logo.

2. Create a horizontal photo 1500x500 that shows your tournament in action. Refrain from adding a web address or any other contact information like phone numbers, etc. Most social media channels have rules against that and may inactivate your account if discovered. Send the artwork to your smartphone.

3. Pick a #hashtag (like #wscsoccer) that you will use across all your social media channels. You are issued a #hashtag on your TourneyCentral site, but you may wish to change that. Search on twitter.com for the #hashtag you wish to use and make sure it is not too popular or used by causes and campaigns that you may not want to be associated or confused with. Putting soccer in your #hashtag is usually a good idea. Keep it short, 5-7 characters.

4. Write a short description of your tournament not to exceed 140 characters. (really, under 140 characters)

5. Set up a gmail.com account as the email account for all your social media channels. (thegamethroughglass@gmail.com)

6. Establish a process for securing passwords and access to each social media account. ***Do not establish under an individual's name.*** Use your tournament and make the access board-level.

7. Download the apps for the social media spaces you plan to use and collect them on one home screen on your smartphone*

**This guide assumes you know how to search and download apps from your iOS/Android appstore. Your phone provider is the best source for this.*

Part Two

Social Media Channels

Pulsecheck

By now, a home screen on your smartphone should look something like this. If not, visit the app store (Apple or Android) and download the apps. I prefer a single screen for all the apps, but you may choose to place them in a folder. Whichever works for you.

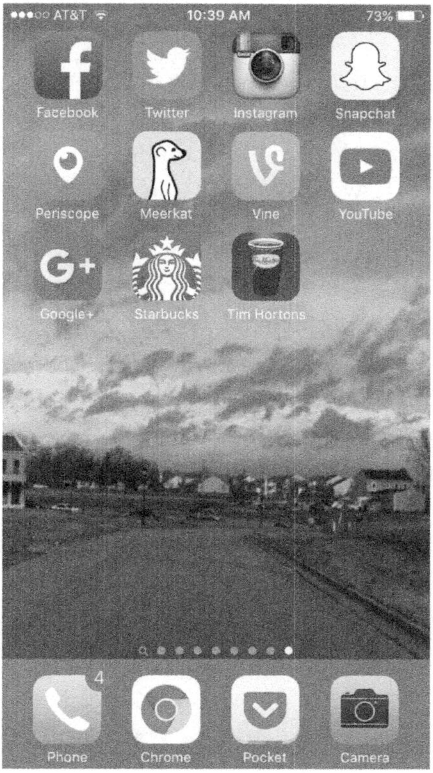

**I added the coffee apps, because you're gonna need them.... Vine and Meerkat died last year... I left the screenshot up to remind you nothing is forever in social media... diversify, be everywhere you can be, build all platforms.*

Facebook

Facebook is becoming the defacto place to organize event information and share content within a circle of friends. While it will never rival the specialization of a soccer tournament website like TourneyCentral for registration, scheduling and scores, it makes sense that your soccer tournament is available on Facebook for your teams and their fans to find and connect up with you there. It is an additional marketing channel you should not ignore.

Facebook strives to be the center of your event. However, you should always make sure the hub of your event is your tournament website. You publish content to your Facebook page in order to draw fans to your tournament site to support your sponsors, advertisers and centralize your communications to the teams to avoid confusion.

Before you set up your facebook PAGE, make sure you are first on Facebook or the person who is administering your page is on Facebook and willing to accept the duties of page administrator. You cannot establish a Facebook page without a personal profile.

While it's possible to set up a page on the Facebook app, it is much easier to set it up on the website. Here's how to set up your Facebook Page.

1. Go to www.facebook.com. Log into your personal profile. Under the setting arrow at the far right in the top bar, select Create Page.

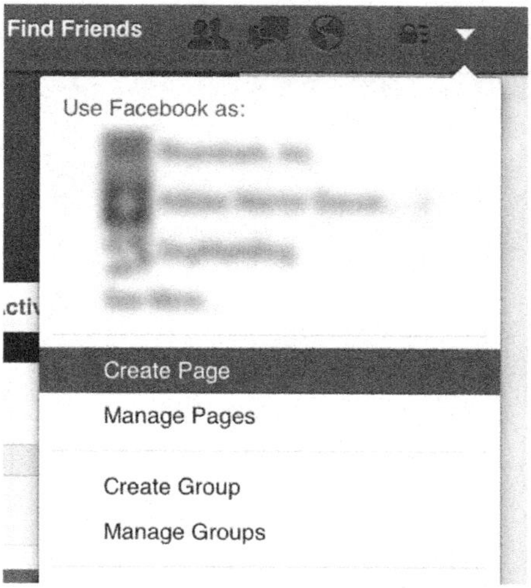

2. Choose either an organization or entertainment. Enter your tournament name.

3. Write a short description of your tournament in 140 characters or fewer in the info section.

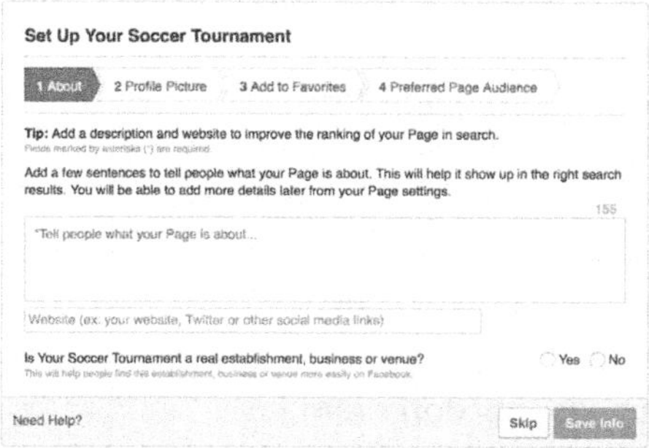

4. You can safely skip the next few update boxes until you get to the last. Then save until you see this.

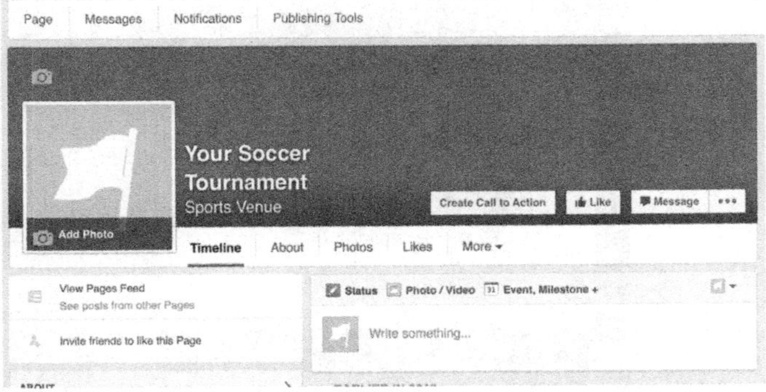

5. Upload your square avatar and cover graphic you created from your prep earlier.

6. Click on the About tab and complete as much information as you can. The more complete your profile, the easier it will be for your teams and their fans to find you.

7. Then click on Settings to the right. These are mostly preferences on how you want to be notified, who you want to see your posts and who can post. For many of these, there are no right or wrong selections, but there are a few you may want to set.
   - Add to your favorites. It will make your page easier to find
   - Make sure your page is published
   - You want to limit the people who can post to your page to your tournament staff
   - You want to set the ability to expire your posts so old announcements don't remain

8. Page Roles > Assign Admins as Editors

9. Be sure to Like your own page.

10. Update your page regularly with the news items that appear on the front page of your soccer tournament web site or photos by posting as your page, not your personal account. Click on the admin triangle you did in Step 1 and select your page to Use Facebook As... Then click on Log In next to your page.

Always verify you are posting as your tournament event and not your personal account when managing content for your tournament event.

11. Start building your fan base by copying the Facebook URL from your page that appears at the top (it will look something like http://www.facebook.com/pages/your-tournament-event/1362877625525) and put that in your Website Maintenance Module>Variables>Go! You will then see a Facebook badge in the left rail of your TourneyCentral website.

Once you have your Facebook account set the way you want, chances are you will be managing the page more from your mobile device that from a laptop.

The facebook app makes it easy to post updates and photos.

1. On the lower menu, click on the more menu

2. Then select the page you wish to manage. We assume only one at this point, but you may have a club page and a second tournament.

3. Select either publish or photo, depending on the content you have.

4. The best way to become comfortable with publishing content to your tournament Facebook page is to just practice posting stuff well in advance of your tournament weekend. You can always delete anything that doesn't look like what you wanted.

   Be sure to assign an expiration date to content that will expire. That is the best way to keep your Facebook page maintenance-free and unambiguous to your teams.

Twitter

Twitter is a service that you can use to fire out quick updates or messages to your followers. This makes it a great tool for youth soccer tournaments and tournament directors because they can immediately announce game scores, special promotions or upcoming events on their smart phones, computers or via text messaging. Twitter — more than Facebook — moves at the speed of a sports tournament.

Like Facebook, Twitter is fastest set up from their website twitter.com, but once established easy and fastest to post using the mobile app.

1. Open a web browser and go to https://twitter.com Click on the Sign Up button in the upper right corner.

2. Complete the first screen with your full tournament name (20 characters max) your email address and password. You can't use the same email on two different twitter accounts.

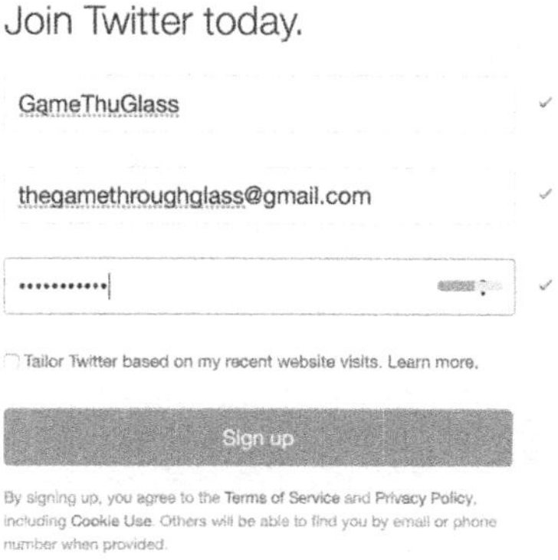

3. You can skip this screen, but you may want to enter your phone number now to make getting text-based messages and set up two-step authentication later.

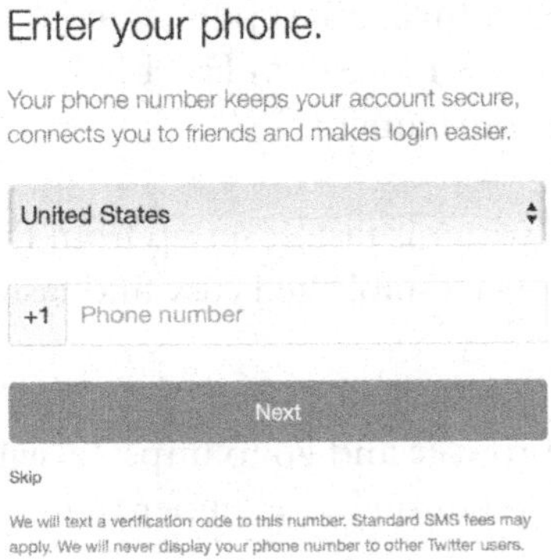

4. Choose a user name. This is different than your real name set up in step 1, but it should be really close. For example, you may abbreviate your user name and spell out the full real name.

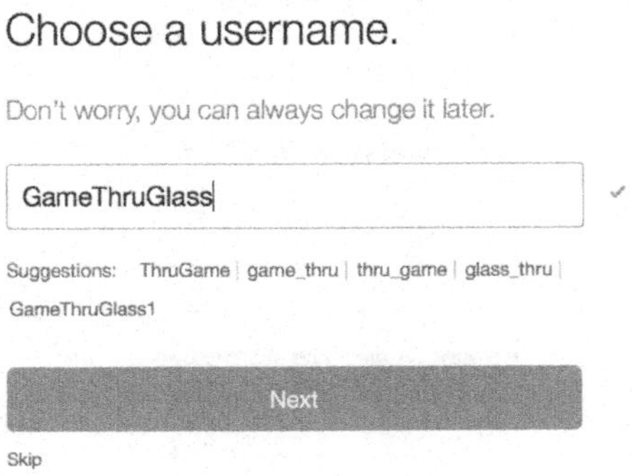

5. Upload your avatar that you prepared in the prep section.

6. Skip through any of the other steps that enable you to follow other users, for now until you get a "ready" screen. Pick up your email and verify your twitter account to get started and remove that pesky reminder.

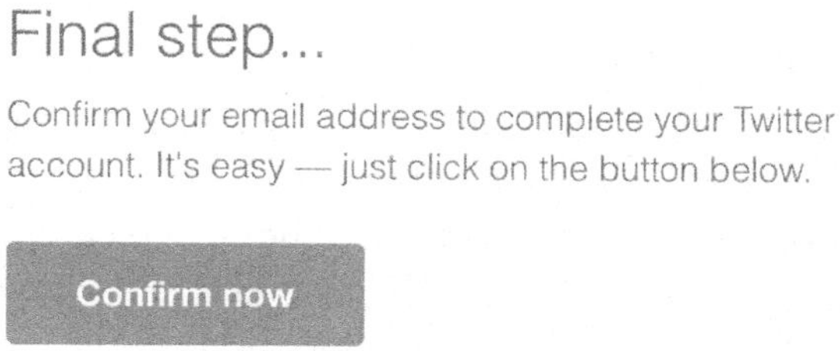

At this point, I recommend logging out and then logging back in to get to a baseline account state.

7. You will also provide a location, your website and a brief bio. From here forward, most of your account settings will be your personal preference, like colors and email notifications. Click through the tabs on the left side and check and uncheck the options. Unlike Facebook, Twitter is very straightforward about what each checkbox means.

8. Fill in your Twitter account in your Web Maintenance>Site Variables set up on your TourneyCentral website. Your Twitter badge will then display on your site so teams and their fans can follow you easily.

Once you have your Twitter account all set up and your options set, we suggest moving over to the mobile app and manage your account from there. On your mobile, you are more agile and quick and able to keep up with the speed at which your tournament moves.

Some mobile basics

Once you log into your Twitter account via the app (the Android version looks slightly different, but everything pretty much works the same) you come to a home screen that lists your tweets and everyone you follow. If you have not followed anyone yet, you'll see your few — or none — depending on how many you sent.

The bottom row is your icons that indicate what kind of notice you are seeing. The Home button is your stream of tweets you see from yourself and people you are following.

The Notifications are tweets that have been sent using your @ handle, people who have recently followed you, liked your tweet or added you to a list.

The Moments is something Twitter is trying out and is not very relevant to your soccer tournament — yet.

The Messages are DM (Direct Messages) your followers can send you that are private... well, as private as the Internet allows. Always assume they are not.

Follow some people

The first accounts you should follow are @tourneycentral and @gamethruglass. We'll then follow you back so things will make more sense and be more interactive.

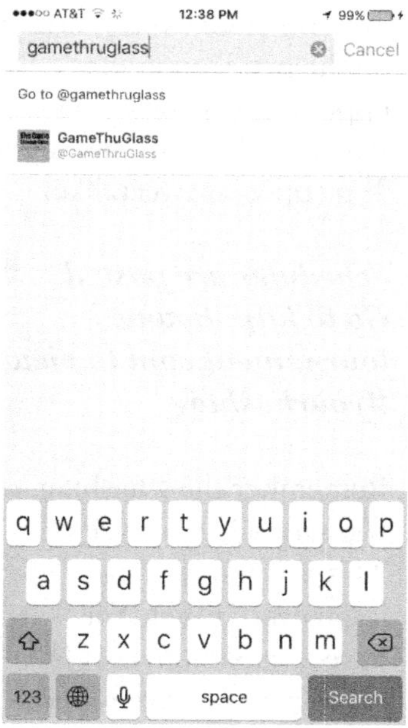

Click on the search icon in the upper right and type gamethruglass. It will appear in the list below. Click on it, then click the Follow button on the screen.

Repeat for tourneycentral.

We will see that you are following and will follow back. Once you see an alert dot on your notification icon in the bottom bar, you will know that we have followed you back. Click the icon and see what it looks like.

Now, go search for other cool accounts to follow back, like your state association and guest teams. Be sure to follow any host club team and request they follow you back.

Sending out tweets

You will probably be using your twitter account mostly to update teams on your tournament. While articles written about Twitter say that you should engage, have conversations, etc, you ultimately decide how you want to use the account. Regardless of what people say, there is no wrong way to do it.

If the only thing you use it for it updating scores, that is perfectly ok.

To send out a tweet, click on the pen icon in the upper right corner. The screen like this will display. Type up your tweet, like:

Schedules are posted. Go to http://www.tournament.com to view #yourhashtag

Remember the #hashtag you set up in your prep? This is where you use it, on every tweet. You have a maximum of 140 characters so leave room for the #hashtag

Experiment with sending out photos on tweets. Click on the camera and search through your photos to attach. (While you can take a photo and video as you are tweeting, we recommend you take the photo/video first, then construct the tweet and attach it.)

Our advice: If you use Twitter for nothing other than sending out scoring updates, that would be ok. You are free to use or not use Twitter as much as you wish.

If you choose to expand out your Twitter use, some ways to use Twitter is to tweet updates before, during and after your event. Any time you post new content on your website, send a quick Tweet out along with a link back to your website.

Your TourneyCentral website is integrated with Twitter. When you update scores, you can send that update notice to your Twitter account. You can also fill in a Twitter account for your advertisers so that a Twitter button appears in their DEALS listing.

Send updates to Twitter during the event, such as sponsor coupons, reminders of lunch specials at concessions, lost and found, etc. Here is an example of what a soccer tournament Twitter account can look like.

The best way to become fluent in Twitter is to use it. Follow some cool folks and watch them talk. Jump in when you feel ready. You can always tweet at us. @tourneycentral

Instagram

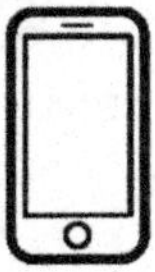

A picture is worth a thousand words. That statement is at the core of what Instagram is. Instagram is one of the quickest ways to share a photo or moment across multiple social media platforms.

It's easy too. For the most part you take a picture, add a caption, then share.

This guide will step your through using Instagram for soccer tournaments. Currently Instagram can only be setup through your mobile device, which is ok because you use your photo to upload photos into your profile.

1. Download Instagram to your phone from either iTunes or Android store

2. Here is where you have a choice. You have the option of signing up with your tournament's Facebook page, or with email. (I suggest using a tournament gmail.com that you have set up instead of clicking the sign up with Facebook button. It is easier to know where your gmail is reaching out to and how things are connected instead of using a secondary social media tool to connect to a third-party app.)

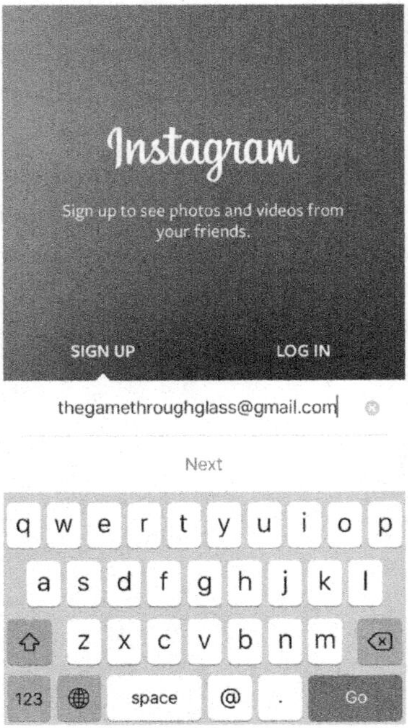

3. Enter a user name and password. I suggest using the same handle as your Twitter account to keep a sense of continuity throughout your social media presence. (If you chose sign up with Facebook account, this is where you will enter you Facebook user name and password.) When you press next, it will ask for your full name and mobile number, both of which are optional. These might make sense for a single person wanting all of their friends to find them according to their phone number, but it really doesn't make sense for a tournament.

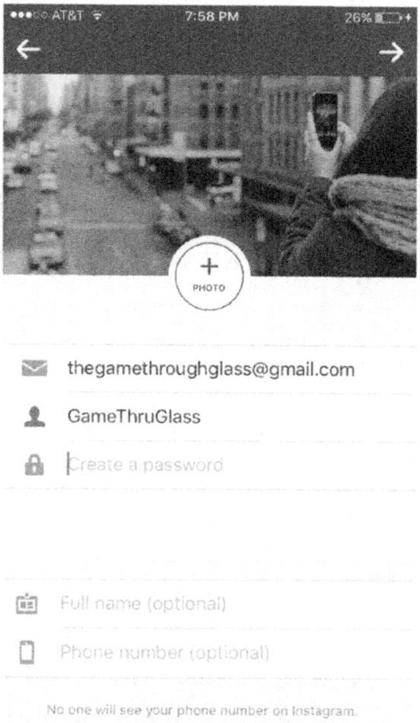

4. Upload your avatar you created in the prep earlier.

5. Edit your profile with some basic information such as your web site. You may want to use the same profile as your Twitter account.

6. The primary reason you want to use Instagram is to post photos of your tournament. So, let's post a photo. On the bottom of the screen is the main menu. The middle button is the photo button.

When you press it you will select either uploading a photo you already have in your phone or take a new one, apply a filter, type up a caption including your #hashtags and post.

7. Here is where Instagram becomes important. It can post to your Twitter and Facebook automatically. In your Settings > Share Settings you will find the ability to link to Facebook and Twitter. Just click on each and it directs you to a page that will ask you to connect the two services. Click ok and every time you take an image, you will be able to click on which services you would like to share the image with. The image also saves to your mobile device. Be sure to enter your Instagram account in your TourneyCentral variables.

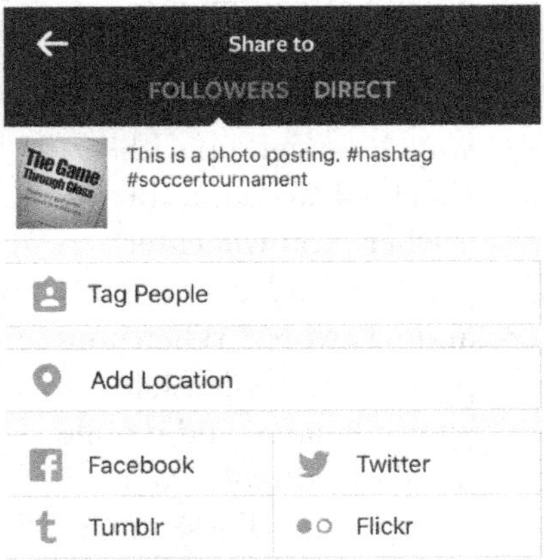

Instagram is really that simple. There are a few key things to remember about sharing photos on Instagram:

Always use a caption to describe what is happening. You aren't there to be "artsy," you are there to build content and experience about your tournament.

Hashtags are important. They are the easiest way for people searching topics to find your photos. #soccer #soccertournament and #yourtournamentname are always good hashtags. If you are featuring your sponsor, you may want to use #sponsorname as well.

What to photograph

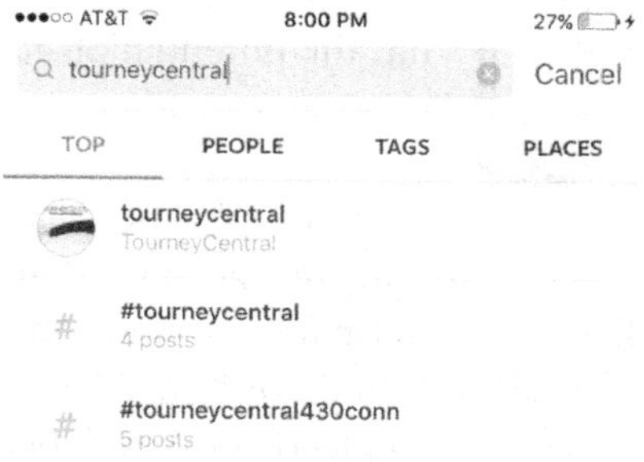

Moments. Any moment can make a great photo. A coach thinking about the next play. A player eating pizza at your registration. A volunteer helping a coach. A referee assignor assigning games. Your scheduler updating scores.

When choosing what to photograph, try to capture the spirit of your event. Showing people having fun doing the most mundane things can give your event a sense of whimsy. Or, if your event is more competitive, show the struggle of putting together the perfect schedule, the perfect event.

Don't forget your sponsors. If you have sponsors at the field, make sure you take lots of photos with their logo shown in the background. Encourage your guest or host teams to photograph themselves at the sponsors place of business. It may be a small gesture, but your eagerness to capture them in your official "record" of photos could mean the difference between a "yes" or a "no" for a sponsorship next year.

Be creative and have fun with photos. But also have a policy and release for kids, especially those under 13. Know the laws and err on the side of caution.

Instagram stories

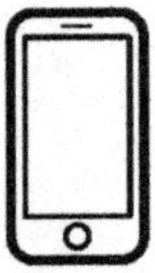

Instagram has added a feature to their platform that has the potential of getting bigger, so we'll cover it as a separate topic — Instagram Stories.

Instagram Stories is a short (10 sec) video clip that you can add to your account that disappears after a day, much like stories in Snapchat. You can decide to do just Instagram or just Snapchat... or both, but whatever you decide to do, make sure you feed your audience on both channels. If you do both, it may be tempting to record on Snapchat, save and then republish on Instagram, but you really should resist and create content unique to each channel for the widest audience engagement.

Regardless of what you do, ***save the video*** before you post it. You will want to create a reel of your social media efforts for your committee, board, sponsors or yourself and it always helps to save before posting.

Getting started

To get started using Instagram Stories, just log into your Instagram account.

1. At the very top of the home screen, you will see a camera icon (or a +) Press that and it turns into a standard recording screen.

2. Hold down the white record button at the bottom of the screen while you are recording your video. Release it when you are done.

3. Add text at the top and doodles as you'd like using the marker and/or text at the upper right (not necessary, but you can experiment as you get more comfortable with Instagram Stories)

4. Save your story to your phone now just in case things go wrong, like you get disconnected or Instagram fails to upload, by clicking on the download icon in the lower right.

5. Send your video to Instagram Stories by clicking on the Record button that now contains a check mark. If you change your mind or wish to re-record, you can always cancel.

6. You can also add photos to your Instagram Stories by just tapping on the Record button once, like you would take a regular Instagram or Snapchat. However, Instagram Stories is turning into more of a short video story tool and the photos are less effective in holding attention. Viewers can see your standard feed for photos.

7. Be sure to also add to your Instagram feed in addition to Instagram Stories. Because Stories go away after a day, you want to have something a bit more permanent. By swiping up on the record screen, you can include any photo or video you recorded on your phone during the past 24 hours. However, people really like seeing original, in-the-moment video so you may want to avoid doing that too much.

What to storify

The key to a good Instagram Story line is to tell a story, not just post up random or sequential videos because you can. That is what you may want to use your regular Instagram feed for.

Have a beginning, middle and an end, even as you may want to tell the story acrosss multiple 10-second clips. Many users string together several recordings to tell a complete story. Always be aware, however, that your videos will drop off at the end of a 24-hour period, so make sure your story still makes sense if the first or second video in a series is no longer available.

If you want to create a multi-video story, it is best to script and set up as much as possible beforehand, so that when you shoot the videos, they are so close in time that they drop off pretty much on the 24-hour expiration.

Have fun and experiment.

And shoot vertical, even though you can shoot horizontal.

Snapchat

Snapchat can increase real-time engagement with your sports tournament by tapping into the real-time energy and your teams' FOMO (Fear Of Missing Out) Despite its beginnings as a teen-aged peer-to-peer network for questionable photos that would "disappear" after being viewed once, it has matured into a solid platform for showcasing a sports tournament.

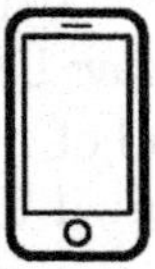

You access Snapchat exclusively through the mobile app. The interface can be confusing and unfamiliar, but this is by design. The kids don't want the olds to be able to use it easily and intuitively. But once you figure it out, you'll be snapping like a pro.

Fortunately, Snapchat has a pretty detailed user guide. We'll run down the basics for you to get you up and running.

The basics

Let's get you up and running. There is really no prep work needed for Snapchat as it assumes you will be using stuff that is happening now in front and behind you. The only think you will need is a unique email address for the tournament that you are not using for any of your personal Snapchat profiles.

1. Download the app and launch it.

2. Press the Sign Up bar at the bottom

3. Fill in your email address, password and a birth date. (you must be over 13) Fill in a user name in the next screen

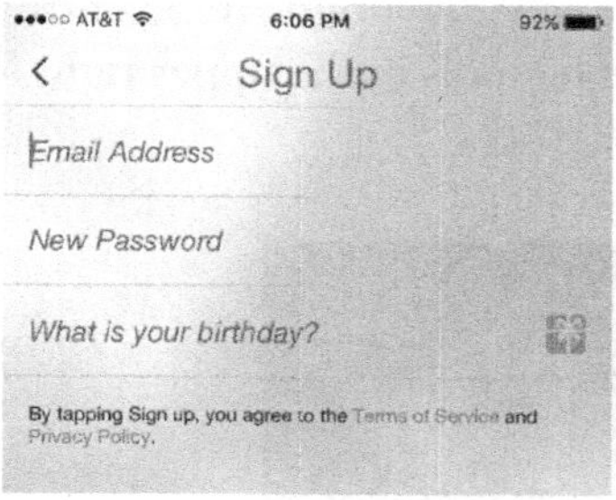

4. Verify your phone number and enter the code they send you so Snapchat knows you're real. You can skip this step if it makes you uncomfortable

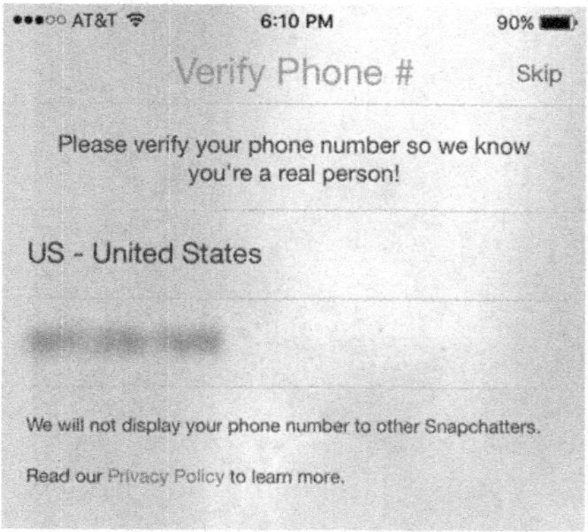

5. Click on the snap ghost at the top, in the settings gear, set who can see your snaps to Everyone under the two settings to maximize your exposure.

5. And you are now ready to send your first "snap." This is the main screen. To create a snap, click once on the button to take a photo. Hold it down to video up to :10 seconds. After you are done, there will be tools in the upper right for text, scribbles and emoji labels.

The best way to figure out the tools is to take snaps and play. They will go away after being viewed and everyone is experimenting anyway, so there are no mistakes.

You can — and should — always add your snaps to your story and send them to your phone for archiving, but it is important to know that snaps disappear after being viewed. You should strive to create each snap as a self-contained mini story that has no history or future. Eventually, your continued snapping will build a story, but each user can and will dip in and out of your stream at will. Remember that and build around it.

Lurk a little

Before you begin to send out snaps, it is best to start following some people first and get to know their style, what they snap, their voice, etc. Here are some snap codes for some snapchatters we think are using the medium effectively. (some language, just so you are aware… this is all real and raw. Also, if you are having issues isolating the snapcode, just visit http://www.tourneycentral.com/snapchat

You can also follow tourneycentral on snapchat, but we will probably just be snapchatting you using snapchat! If we see you follow us, we will follow back.

1. To add a snapchatter by snapcode:
   The page snapchat.com/support shows you how to add by snapcode. Basically, login, point your camera at the snapcode and hold down on the screen.

2. Watch the snapchatters for a few days to get a sense of their style, what you like and what you think you can adapt to your tournament event. You access their snaps by pressing on the purple square at the lower right corner of your login screen. If there are no snaps, you will see a "hamburger menu" with three lines.

Are you getting the hang of the interface? It is a bit different — as a reminder, this is by design to frustrate older folks away from the app — but press on and the connection with your tournament audience will be incredible.

As you watch the snapchatters above, you will see they use filters and geofilters (location/event overlays) video, slo-mo, reverse video, text and drawn overlays. When you find yourself asking, "Hey, how did they do that?" navigate over to snapchat.com/support and find out.

3. Get brave and start looking for folks who snapchat soccer and start adding them. Be sure to read the next chapter about building out your social media networks.

4. Publish your snapcode in a news item on the front page or as a sponsor on the sponsor page.

What to snapchat

By now, you should be ready to just dig in and start snapping. What to snap, what to snap?

1. Your snaps are a part of your tournament event and must be consistent with your brand. Will you use just one spokesperson on snapchat or will there be a team? Will you "tag team" the snaps or have pre-planned programming hours where one person hosts a block of time? Will you invite guest teams to be "interviewed" on snapchat? There are no right answers, but you should think about this ahead of time.

2. Don't be too programmed. Nobody likes being marketed to, especially on snapchat. If your audience senses you are "playing to a script," they will drop you quickly. Just be yourself — or at least the self that is the tournament brand.

3. Design a day. Tell mini-stories about your tournament as the day progresses, before during and after the tournament. Show what is it like to organize the paperwork for registration, schedule games and referees. Take snapchat out to the fields on prep day and show the field lining, the nets going up. Interview the volunteers. Have a sneak peek at the tshirts at the HQ tent. Show the picking process for the shirt pre-orders, the late-night pizza for the hard-working volunteers.

4. Show the activity behind the scenes. Your guest teams want to be invested in your success as a shared experience. Give them that, let them in on the know.

5. Invite your advertisers, vendors and sponsors to run a promotion on your snapchat stream. It could be as simple as producing a short video in real time with the vendor of a popcorn stand at the venue saying "Stop at the Kettle Korn stand at Grantham Park, show us the next snap with the promo code and get $1.00 off" then create a photo snap of the vendor's logo from their booth with the promo code handwritten across it.

6. But mostly, have fun!

Snapchat supports URLs and your TourneyCentral site supports them as well. Just add your Snapchat account in your Website Maint Module>Variables and it will show up on the side rail and in your mobile app.

Periscope

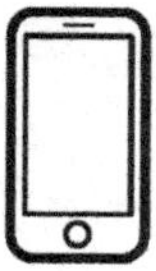

Periscope is an app that turns any smartphone into a broadcast video camera. Basically, you download the app, log in with your twitter account and press the stream button to become a live broadcast station. You put in a title for your streaming session, point your smartphone at the action, and click on a button to start streaming. Your followers will get a notification on their twitter timeline that you are streaming and they can watch and comment along with you. When you are done, you can save the stream to your phone.

There are other features like flipping the cameras around to point toward you, tilting the camera from vertical to horizontal, a chat button that functions like twitter as well as a "hearts" button viewers can use to give you likes. Once you are done streaming, your session is available for the next 24 hours and then disappears. Like Snapchat, Periscope is designed to increase the real-time engagement with your audience.

1. Launch the app and click on Log In with Twitter. Select your twitter account. (you need a twitter account to use Periscope.)

2. Periscope will fetch your twitter account name and avatar. You should use the same avatar as you twitter account and most likely, the name you use with your twitter account will also be available for your Periscope account. If it is not, Periscope will suggest an alternative. You should accept its recommendations and click on Create Account.

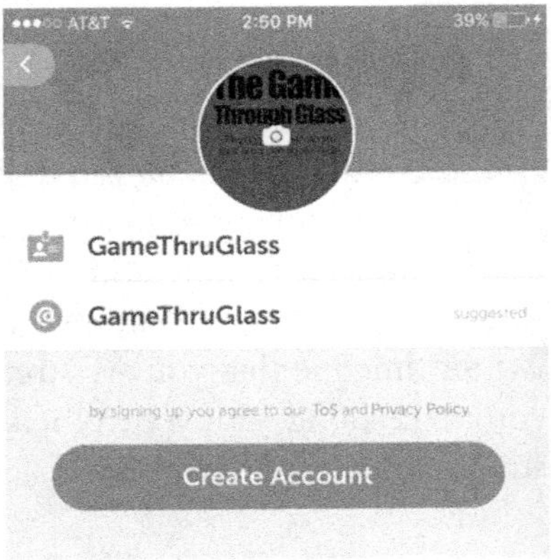

3. Periscope will present a list of other accounts you may want to follow. Click on the follow icon next to each if you'd like.

4. In the upper right corner, you have a button for your settings. Click that and toggle Autosave Broadcasts. This saves you the extra step of remembering to save the streams to your phone. If you have a smartphone with limited memory, you can keep this off and save

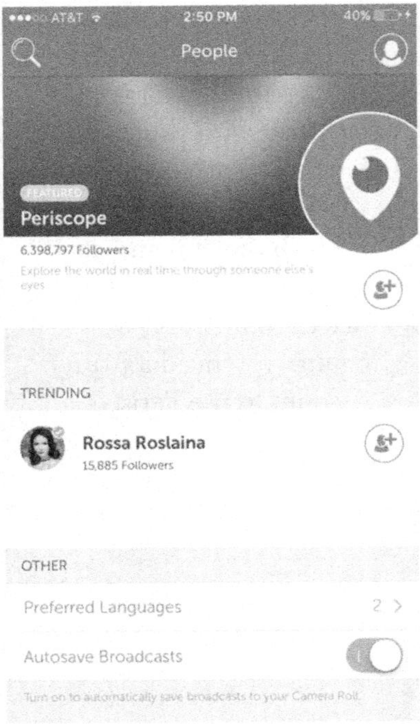

selected streams to your phone. You may also choose to live broadcast in short bursts instead of long sessions. For example, you can stream the kick off or coin toss and not an entire game.

5. When you are ready to broadcast, click on the Broadcast button in the bottom menu bar.

The Start Broadcast screen will pop up and the camera will focus on what you are broadcasting. Make sure at this point you are showing what you want to show. Also, the mic will be live! Enter a name for your session, toggle the geolocation on/off (we recommend on during the event at a public park, off for any other use) and the twitter icon to send notifications out to twitter. When you are ready, click on Start Broadcast.

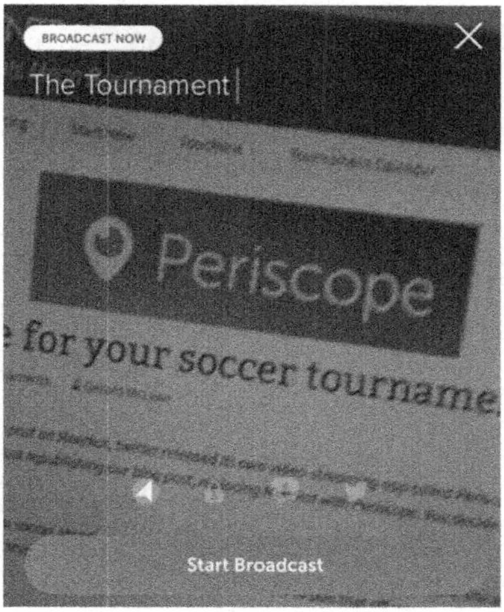

6. **How do I stop broadcasting?** This is most asked question about Periscope, mostly by the camera operator, live during the broadcast stream. Swipe Down from the top, then click on Stop Broadcast. To toggle the front and back cameras, just double-tap the screen. Like using a speaker phone, make sure the stream is 100% done before saying anything out of turn or colorful. It is Internet and it is live!

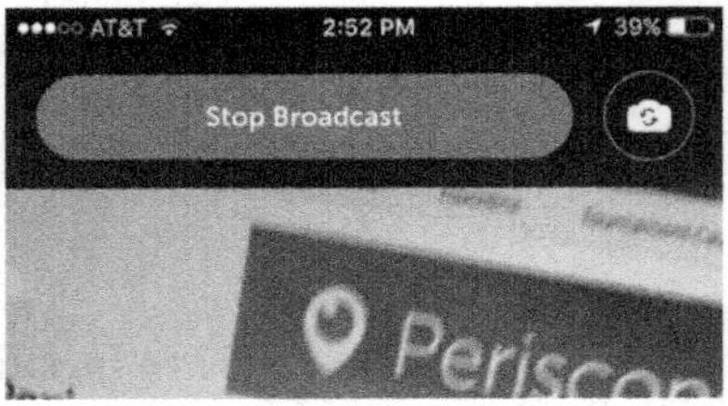

Since Periscope relies on a strong Twitter network, you should focus on building your Twitter following. Running up to your tournament, set up some behind-the-scenes Periscope sessions such as a welcome message from the tournament director or a Q&A session where you encourage questions be submitted live via Periscope. Periscope is also live directly in a user's Twitter stream. No app needed to view the livestream.

You will see viewers come into your stream as you first start it. Give it a minute before launching into the show and welcome the viewers as you see them come in. Let them know they are one of the first to arrive and make sure they don't click away because you are "boring."

Visit an advertiser/sponsor and Periscope the experience. Ask the owner/manager to invite the teams or welcome them. Do this before, during and after the tournament.

Invite parents to live-cast the games during the entire weekend. Post a large banner with the Periscope logo and a white space where anyone willing to live-cast can put their Twitter handle so fans and players can connect and watch each others' Periscope. Convert this into a front page, sticky story so people can find and follow the live video streams from your tournament.

Assign someone from your tournament to be the official Periscope stream from the tournament. Video games, the HQ, interview vendors, show the action, show the fun. Show the teams that decided to go to another tournament instead of yours how much fun they are missing!

Practice before the tournament.

I'm sure there are all sorts of other ways you can use Periscope for your event. Feel free to share them with us and we'll post your creativity on our channels.

You may decide to also use Meerkat in addition to or instead of Periscope. With a few minor difference, the apps work pretty much the same way.

YouTube

For our purposes, YouTube is a container of targeted, scripted and edited video messages. While a standard video used to be five minutes or longer, now it is more common for video to be trailer length of no longer than three minutes.

Because YouTube is part of Google, signing up is easy. Just log into your gmail account you set up in the prep earlier, locate the apps grid at the top and select the YouTube icon to activate your account.

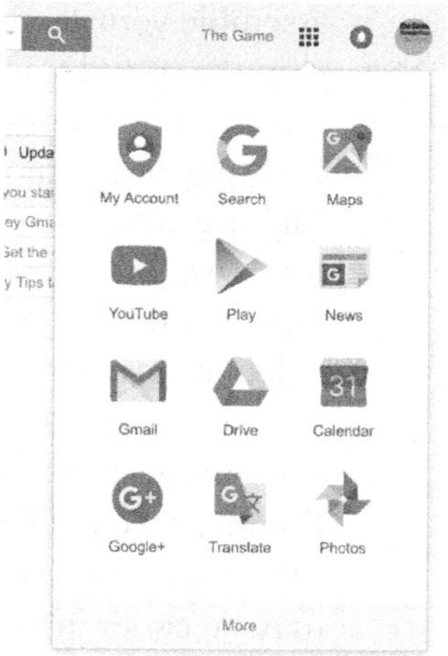

The next screen will enable you to upload a new avatar if different from your gmail account (probably not) and to confirm your birthdate. You can then upload videos created with video editing software or shot on your phone, etc.

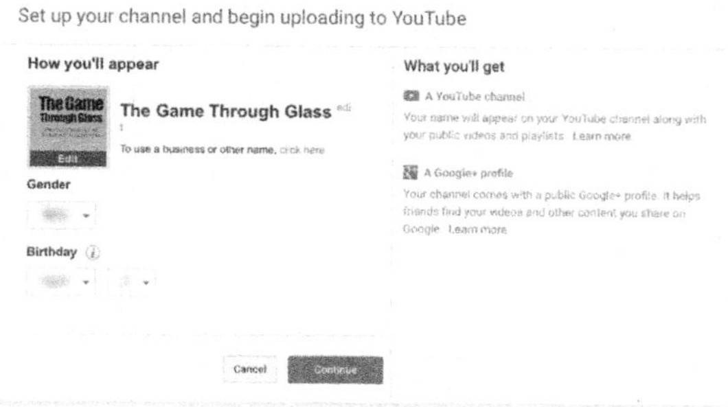

There are some basic rules for using YouTube in your sports tournament:

- Shorter is better. Keep promotional videos under a minute if you can.
- When producing a "How To" video such as how to assemble your team registration packet, script it. Then strip the sound track out and do a voice-over to tighten it up even further.
- Lighting and audio
- Lighting and audio (duplicating this step is not a typo; it is important)
- Hire a good editor. You can shoot and edit a YouTube video on a smartphone, but we really don't recommend it. The app is really only good for viewing the video.
- Edit on the desktop using good editing software like Premier or Final Cut Pro, even if you shoot with a smartphone. Avoid consumer software such as iMovie. It may be easier at first, but you will find it limits your skill and credibility.

Google Plus

The latest player in the suite of tools known as social media is Google Plus, or more commonly seen as G+ in buttons. While it can't boast anywhere near the number of users Facebook and Twitter have, it has one advantage the others don't — Google Search.

Most Internet users start off a web session on google.com, whether or not they know the direct web address of the site they are looking for. Many simply type in keywords like "soccer tournaments in Ohio" or "soccer tournament software." If these keywords are in your G+ posts, you have a greater chance of being found at or near the top of a search result.

If you set up your YouTube account, you already have a G+ page. Just log into your gmail account, click on the apps grid and select G+.

1. Upload your cover photo you created in the prep earlier.

2. Click on the About tab and fill in the details about your tournament, including a location so Google Maps can plot your venues. The more details you provide, the better Google will be able to index your tournament.

3. Include your G+ page in your TourneyCentral variables to make the G+ icon appear on your website.

That's pretty much it for setting up your G+

Our Advice: Posting to G+ is similar to posting on Facebook so you should be immediately familiar with most of the interface. Repost your front page news, scores, sponsor offers and photos to your page. Don't worry so much about interacting on your page as a social media channel as your primary purpose is to push keywords into Google search.

Be sure to include keywords in your post and photo captions as these keywords are pushed directly into the Google search engine.

You can follow TourneyCentral on G+ and we will follow you back. However, we use our G+ page pretty much for keywords too and not as a social engagement tool.

Part Three

Putting it together

Building out your network

You have all the tools to participate fully in social media for your soccer tournament. Now, let's put this all together and start building out your tournament social media network.

Twitter, Facebook, Snapchat accounts from all your coaches. Follow all your coaches on the various channels and encourage them to follow back.

Comment on their postings. The more your coaches feel connected to your event, the more they will want to participate.

Draw boundaries. If you will not reply to game change requests, referee dissent, etc on the social media channels, let them know that, but give them the proper channel such as your contact page on your website. Don't just leave them hanging.

Parent twitter and facebook accounts

Encourage the coaches to have their parents follow the tournament channels early, especially twitter and facebook if you will be sending scores out that way. Few things motivate parents to follow your tournament that knowing they will be receiving scores there. Be sure to place a disclaimer on any follow request that only 13+ should follow back.

Do not follow parents back. Your relationship is ultimately with the head coach. If you follow a parent back, they may take that as an invitation to engage with the tournament directly.

State associations

Follow your state association. Congratulate achievements as you see them.

Sponsors and Advertisers

Follow your current AND prospective sponsors and advertisers on all your channels. When they are having a promotion, comment (but don't stalk them.)

Frequent their establishment as a customer and comment on your experience. Share this list with your coaches and team parents and encourage them to do the same. Be a good citizen and they will want to support your soccer tournament.

Prior to the tournament weekend

Make sure you send out and post your social media activity. If you will be doing a live Facebook or Periscope show from the HQ roving golf cart, let teams know this. Ask that they follow you and watch our for the cart as you might capture them warming up, playing their game or generally having a fun time. Let them know you will be stopping often and conducting ad hoc "interviews."

During the event

Yes, you will be busy but you can't use that as an excuse to not engage in social media. If you have prepared properly, you will have a community team in place already whose only job is to engage with your teams on social media channels.

Make sure the social media team knows the boundaries you have set. It is not "just twitter" or "just facebook" anymore; anything said on your social media channels is done as an agent of the tournament.

Make sure the social media team has a robust data plan. Few things are a buzzkill like running out of data halfway through the first day.

Plan for an all-day Periscope live broadcast.

Whatever happens on the fields, happens in your social media channel. You may also encourage local TV stations to tap into the stream and use appropriate footage to showcase your tournament.

Send up a steady stream of snaps to snapchat. Also, scores, advertiser updates, coupons, etc.

The trick to using social media during the live event is to select the tools you can support and pace yourself, but you should be in several channels. Don't put all your effort into just Facebook.

However, you do not want the social media team to become too ambitious and become overwhelmed the first day. The first day will also be the lightest, with a buildup as teams discover your activity. Prepare for a surge in @, replies, RTs, reposts and questions.

Display the #hashtag everywhere, on the golf carts, at the HQ, on the staff shirts, at the concession stands, on the field signs.

Encourage parents and players (U13+ only, please) to post their photos with the hashtag.

Your social media team should be constantly searching for your hashtag and/or event and pulling in photos, tweets and posts, commenting and RTing, etc.

After the event

If you have done a good job during the event, you will see some residual traffic over the next few days as your posts and team/player/parent posts are shared. Keep engaging! You may well be tired, but push through and stay as real-time as possible.

Dormant time

There is no dormant time. Soccer tournament marketing is a 365/24/7 endeavor. Re-read this chapter and start all over again for next year as soon as you can.

A common theme throughout using social media is be a good citizen. Follow back, engage lightly and appropriately, don't sell, interact. Be human. Never get comfortable with your follower count on one platform… what's hot this year may be nonexistent next year. You want a healthy cross-over audience on each channel. Be prepared to switch focus as a new social media channel emerges (don't worry about staying current… just follow TourneyCentral for the latest social media trends.)

Follow TourneyCentral on our channels. Chances are, if we are doing something cool, you'll want to get in early.

Part Four

Tools

Media Planner

A sports tournament can be practically divided into about eight phases. When planning your social media, it is important to understand what phase the event is in in order to craft the message properly. For example, the teams are not going to care about the deals from the vendors/sponsors if they don't know yet if they have been accepted or when they will play.

Here are the eight phases and what the tournament and teams are doing during this time. As part of the planning, you should determine which social media channel to use and what the message should be that best satisfies both needs.

Keep in mind that during the tournament play, the scores will also most likely be sent out in burst from the scoring teams to Facebook and Twitter. While the teams may tolerate a greater activity rate during those times, you should be aware of potential messaging fatigue and plan accordingly.

Lastly, while the tournament and social media team is broken out separately in the matrix that follows, it is only for purposes of this guide. The tournament director and social media team work very much in concert. Your event playing out on social media is not separate from the physical event, so there should be tight coordination between the two.

Phase I – Latency

Tournament:

- Make sure all information is up to date and reflects the status of the tournament. Get your sanctioning, put together a marketing plan and begin writing your broadcasts.

Guest Teams:

- Guest teams must likely are not thinking about your tournament as much as you are. Don't worry; they will.

Social Media Team:

- Write and post updates about the tournament planning progress.
- Include key activities such as hosting approval, deadlines, costs.
- Highlight past teams and clubs.

Phase II – Marketing

Tournament:

- Broadcast email messages to last year's teams, let them know you are there for them.
- Prepare and mail postcard marketing.
- Make sure your link on the association web site is up and working.
- Attend divisioning meetings and hand out marketing materials to potential guest teams.

Guest Teams:

- Guest teams are looking at the website and deciding which tournaments they will be applying to. Teams are also looking at past team scores and standings.

Social Media Team:

- Pre-write some tweets and Facebook posts
- Collect photos from last year; prep as illustrations to posts
- Write some blog posts for the website.

Phase III – Applications

Tournament:

- Applications will be coming in. As they come in, you should look at them and begin to “seed” them into large groups (1, 5, 9, etc) Don’t worry about seeding precisely.
- Front page breaking news, quick news and the scroll message should be updated to reflect the application phase.
- Teams should be marked as they pay. Messages should be going out to teams who have not yet sent in their check within the past 5-7 days.
- Teams who apply should be getting information email messages at least every two weeks, to keep them informed.
- FAQ postings accurate and timely. Prepare FAQs for registration, etc.
- Forms for registration should be prepared and finalized, including the shirt design.

Guest Teams:

- Guest teams are logging into their application and keeping track of their fees being paid, rosters received, travel, etc.
- Guest teams should be encouraged to keep their information accurate and up-to-date.
- Guest teams will be able to change their application until the deadline, after which they will be able to only see their status.
- After the deadline, all applications will be marked Waiting, in green.

Social Media Team:

- Read and respond to FAQa in Facebook posts, tweets and snaps
- Use Instagram to illustrate a tournament team in hightened readiness

Phase IV – Acceptance and Confirmation

Tournament:

- Turn off the application FAQs and turn on the registration and confirmation ones.
- Check all the teams that you are accepting in one action using the "mark" tool. The guest teams are monitoring their status almost every hour.
- Make sure that ALL your registration forms have been placed on the web site.
- Change the front page, Quick news.
- Send an email to the guest teams letting them know they have been accepted and they must CONFIRM by logging into their application.

Guest Teams:

- Guest teams will log in using their TEAM ID NO that appears on the top of every email that is sent from the Applications Module.
- Teams will log in, click on the "Click Here to Confirm" link and their application will be tagged as Confirmed and the date and time will be written as part of their record.
- A menu of forms will then be presented to the team.

Social Media Team:

- Continue to assist the tournament staff on making sure confirmation compliance is high
- Highlight sponsors and DEALS to incoming teams
- Assist in Facebook posts to ensure teams know how to get forms for registration.

Phase V – Scheduling

Tournament:

- Build the list of tournament divisions. If you don't know all the divisions, build at least one per age group.
- Build a list of winner placeholders. Be sure to include Wild Cards and qtr finals and semi final winners, if necessary.
- Make final decisions on the team seeds and mark accordingly.
- Pair teams with the scheduling tool. If you use an off-line scheduling system, be sure that your on-line schedule matches. When you are done, publish your schedule. Be sure to update the schedule using the schedule tool if things change.
- Broadcast that the schedule has been posted. Change your front page, quick news to reflect your status.
- Email broadcasts should be going out regularly, but not too often as to create fatigue. During the last 2 weeks, they should be going out at least once a week. The week leading up to the tournament is your biggest pre-event traffic week, with heavy hits on restaurants and hotels.
- Write your THANK YOU MESSAGE NOW!! You will be too tired to think of anything on that last day.

Guest Teams:

- Guest teams will pick up their schedule from the website.
- Registration materials completed.
- Teams are checking out places to eat, entertainment, especially during the last week leading up to your tournament.

Social Media Team:

- Read and respond to FAQs in Facebook posts, tweets and snaps
- Mine the schedule for media hooks. Pre-interview teams/clubs for interesting personal stories.

Phase VI – Registration

Tournament:

- Create and communicate an online registration program.
- Receive and approve registration materials.

Guest Teams:

- Teams prepare and send their official registration paperwork.

Social Media Team:

- Continue to assist the tournament staff on making sure registration compliance is high
- Report on the progress of the registration. If your tournament event has a physical registration party, find those individual stories and report with photos, videos and posts.
- Mine the team rosters for active social users to connect with during the tournament play

Phase VII – Tournament Play

Tournament:

- Scores updated through the admin section.
- As winners emerge, change the team pairings on the schedule.
- Update the front page, quick news to reflect the current action. Plug vendors at the tournament, etc. People are visiting the web site during the event.

Guest Teams:

- Guest teams are checking out scores, DEALS, information FROM THE SIDELINES as the games are going on for updated information.

Social Media Team:

- Make sure the scores are flowing to the social channels
- Take a lot of photos and videos from the fields and provide a constant flow to the social channels
- Monitor the social channels and report to the tournament scores discrepencies, complaints, etc. Make sure you have a response plan in place.
- Conduct ad hoc interviews on Snapchat, Periscope from the field or HQ with teams, sponsors and staff
- Plan for weather! Social is the first line of communication for inclement weather.

Phase VIII – Post Tournament

Tournament:

- Post your thank you message up on the front page.
- AUDIT your scores. Make sure they are complete and accurate.
- Take care of the lost and found items. Mention the lost and found page in your final broadcast.
- Move into Phase I for next year

Guest Teams:

- Teams will get home, check out scores until about 2-3 days after the tournament.

Social Media Team:

- Post lost and found items in the social channels
- Recap from winners and finalists
- Follow up with any last minute concerns from the teams
- Analyze the best media pieces and plan for next season

Whole event planner

Phase / Channel	Facebook	Twitter	Instagram	Snapchat	Periscope	YouTube	GooglePlus
Latency							
Marketing							
Apps							
A&C							
Sched							
Reg							
Play (Day One)							
(Day Two)							
Post							

Single event media budget worksheet

This form is available as an online database in Game Through Glass webinar series.

Event Name: ______________________________

Date/Time: ______________________________

Location: ______________________________

☐ Must Get:

Contact: ______________________________

Channel(s):

☐ Facebook ☐ Twitter ☐ Instagram ☐ Snapchat

☐ Periscope ☐ YouTube ☐ G+

Media:

☐ Photo ☐ Video ☐ Text

Format:

☐ Promo ☐ Action Shots ☐ B-Roll ☐ Interview ☐ Stream

Description:

Tweet/Post: ☐ Scheduled ________________

Passwords

Account	User	Password
Facebook		
Twitter		
Instagram		
Snapchat		
YouTube		
Google+		

Part Five

Advanced Social Media

The 8:00am rule

If you have gotten to this page, you have probably read through this entire book, combing for pearls of wisdom that will give you a competitive advantage over events that are merely using social media tools because they feel they have to. I feel I should reward that tenacity with a secret, so here is my most guarded one.

When we used to shoot action photos at TourneyCentral, I had an iron-clad rule. All photographers were to be at the fields at 7:45 am, have 20-30 shots on the card and the rest before half-time. They would then be required to return the cards to the production van and everything would be processed and linked up to the games before the game scores were posted on the website.

The tournament director was also called and reminded that the 8:00am round of scores should be posted within 15 minutes of the games being completed. We would even do it for them if needed.

Why? Simple. People were told the website was real-time, but years of paper and confusion made them skeptical. We were there to prove them wrong and set a higher expectation for a TourneyCentral website. And give our competitors a bit of heartburn along the way, but mostly to be exceptional.

If parents logged into the soccer tournament website and saw the photos were being posted even as the games were being played and the scores up within minutes of the games being completed, they got a sense that everything was buttoned-up and real-time. They would be more likely to return repeatedly throughout the day to check scores, sponsor ads and photos.

However, if they did not see photos and scores for the first round of games, they would shrug and assume that things would "eventually" get updated and resolve to maybe check at the end of the day. A soccer tournament is a busy place.

When we analyzed traffic at the end of the day, we would consistently produce 300-400% more than events who opted out of our photography services and instead went the traditional sub-contracted route with other photog crews.

But here is the funny little secret: We did not need to keep up the pace all day. We could then shoot several rounds at a time before returning and processing the photos. The parents and coaches never knew we didn't keep up in real-time all weekend. For them, nailing that first round set their reality of us being real-time. We just never argued with them.

If you are running an event, like a sports tournament, always nail that first round. It sets the tone and expectation for the rest of the event. Instead of playing catch up the whole event, you'll have your audience swearing that everything happened in real-time online, even though it may not have entirely.

Social media updates increases that excitement of being at the event. Your guest teams will actively participate if they see you care and are connected. You will also be connecting families who could not be at the games because of travel, health or other commitments.

Social media as a jobs training program

The core skills that every information-age worker needs today to compete in the modern economy are; photography, videography, writing and social media.

Playing soccer on a team in a competitive league gives young people critical skills that will serve them well throughout their lives. A smart tournament can also give young adults an opportunity to hone critical job-market skills by opening up their social media accounts to them to "practice" skills they will need when they search for a job.

Because there are many different channels within social media, the opportunity to bring on many people to manage different channels is almost infinite. Of course, like any good team, you will need a coach to coordinate, monitor and guide your media team.

Social media has a ton of potential to add to your tournament brand but it is deceptively a lot of work. It is intense and fast-paced but you have the potential to give someone one heckuva resume reel!

If you do it right, your tournament can be a competitive jobs program where the best media minds can not only learn new skills but give back a deep, rich online space for your tournament event.

Basic Google Analytics

As a social media producer, you've convinced your tournament that it needs to be on social media. You've had a lot of fun. You've worked hard at grabbing photos, tweeting, got some cool video and a few interviews with sponsors.

Now the club wants you to prove an ROI. Here's the bad news; you can't. Not directly because social media isn't its own thing — it is symbiotic with the action on the field.

You can get caught up in measuring social impact, where traffic is coming from, etc., but if you are producing like the whole world is watching and you are getting engagement from your teams, that's all you really need. Eventually, deep social will take hold and will become expected from every tournament.

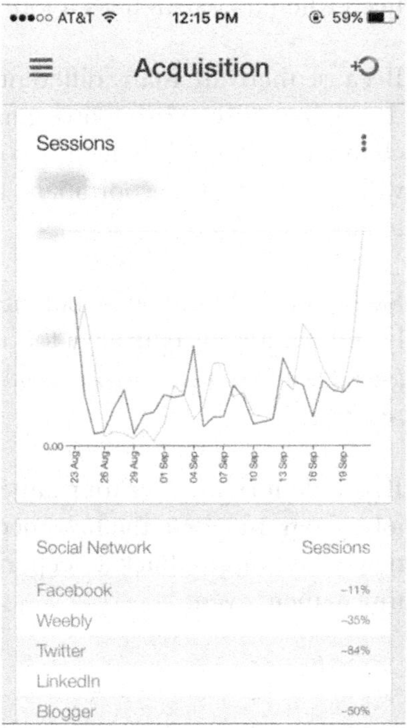

But you will want to keep track of activity anyway and the Google Analytics app is a good way to do it. Once you set up your Analytics account with your gmail address, download the app, log in and watch traffic and the social engagement.

Typical traffic pattern

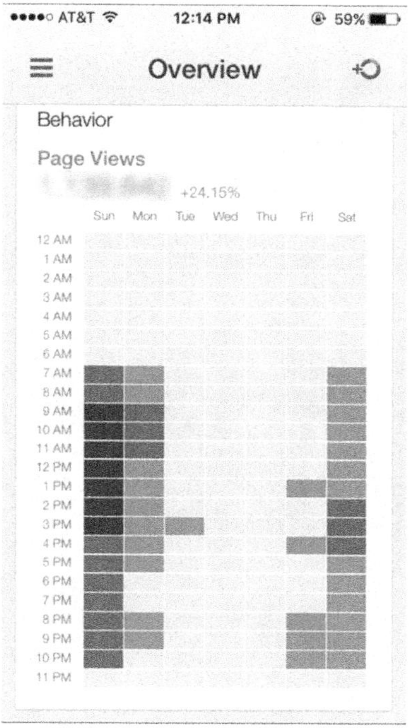

The traffic pattern is pretty much what you would expect to see, with heavy weekend traffic and significant flow when the scores are updated.

Remember our 90 Minute Attention Span on page 19? Your job is not to work the darker blue slots (they get there organically) but to make the light blue slots darker blue and the white slots light blue. How you do that is a combination of the skills you learned with this guide, your intuition and curiosity.

There is more

This is always more, but those are more advanced skills than can be covered by this guide. At TourneyCentral, we are always looking for way to expand the sports tournament experience and crafted a webinar series to expand this guide.

Visit us at **ww.thegamethroughglass.com** to learn more and sign up for our value-added services.

Secret code: **RTIGTG**

About the author

Gerard McLean is the founder of TourneyCentral, a full-stack sports tournament software package that manages an event online from marketing through scores and standings. In 1999,

he pioneered the concept of real-time photos and media linked directly to a sports schedule and has continued to push the envelope of integrated media. With experience in retail merchandising, CBT training systems, video production, educational curriculum and newspaper design, he continues to challenge the boundaries of digital media. McLean is also a prolific writer and is a published author of several collections of essays.

Made in the USA
Monee, IL
04 November 2026

67377183R00056